EARTH'S LANDFORMS

ISLANDS

by Lisa J. Amstutz

PEBBLE
a capstone imprint

Pebble Emerge is published by Pebble, an imprint of Capstone.
1710 Roe Crest Drive
North Mankato, Minnesota 56003
www.capstonepub.com

Library of Congress Cataloging-in-Publication data is available on the Library of Congress website.
ISBN: 978-1-9771-2456-2 (library binding)
ISBN: 978-1-9771-2632-0 (paperback)
ISBN: 978-1-9771-2499-9 (eBook PDF)

Summary: Describes islands, including their features, how they are made, and where they can be found around the world.

Image Credits
Capstone Press/Karon Dubke, 20; Dreamstime/Lerka555, 19; iStockphoto: Gfed, 5, Zoran_Photo, 13; Shutterstock: Artsiom Petrushenka, 15, BakerJarvis, 17, Blue Planet Studio, 18, Delpixel, 11, Helen Hotson, 7, Huw Penson, cover, iAmMrBenjamin, 10, Kasia_Przygodzka, 6, Maridav, 14, Parshina Marina, back cover, Pitroviz, 12, Yvonne Baur, 9

Editorial Credits
Editor: Alesha Sullivan; Designers: Elyse White and Bobbie Nuytten; Media Researcher: Kelly Garvin; Production Specialist: Tori Abraham

All internet sites appearing in back matter were available and accurate when this book was sent to press.

Printed and bound in China.
003322

Table of Contents

Words in **bold** are in the glossary.

WHAT IS AN ISLAND?

Have you ever seen an island? An island is a piece of land. It has water all around it.

Islands are one kind of **landform**. Islands come in all sizes and shapes. Some islands form as Earth changes. People make other islands.

HOW DO ISLANDS FORM?

Some islands came to be long ago. Earth had one large piece of land. The large piece broke into smaller pieces. Some of the smaller pieces became islands.

Later, ice covered much of Earth. When the ice melted, it turned to water. Some of the high pieces of land were still above the water. They became islands.

Volcanoes can make islands too. Some are in the ocean. Hot liquid called lava flows out of them. It cools and turns to rock. The rock gets higher each time this happens. Layers build up. Then they push above the water. They make an island.

lava
volcano

A small piece of land can become an island. Water and wind break down the land. Over time, the land **erodes**. An island is left behind.

Sometimes people make islands. They pile up rocks and sand in the water. Then they can put up buildings on this new land.

WHERE ARE ISLANDS FOUND?

You can find an island anywhere there is water. Some are in rivers and lakes. Some are in the ocean.

Sand or rocks can pile up in the sea. They can make an island. Water in the sea splashes on the island. The island keeps waves from hitting the **mainland**.

BIG AND SMALL

Some islands are very large. The biggest island is the country of Greenland.

Most of the country is covered with ice. Some of the **coasts** have big **icebergs**. Greenland also has high mountains.

mountains in Greenland

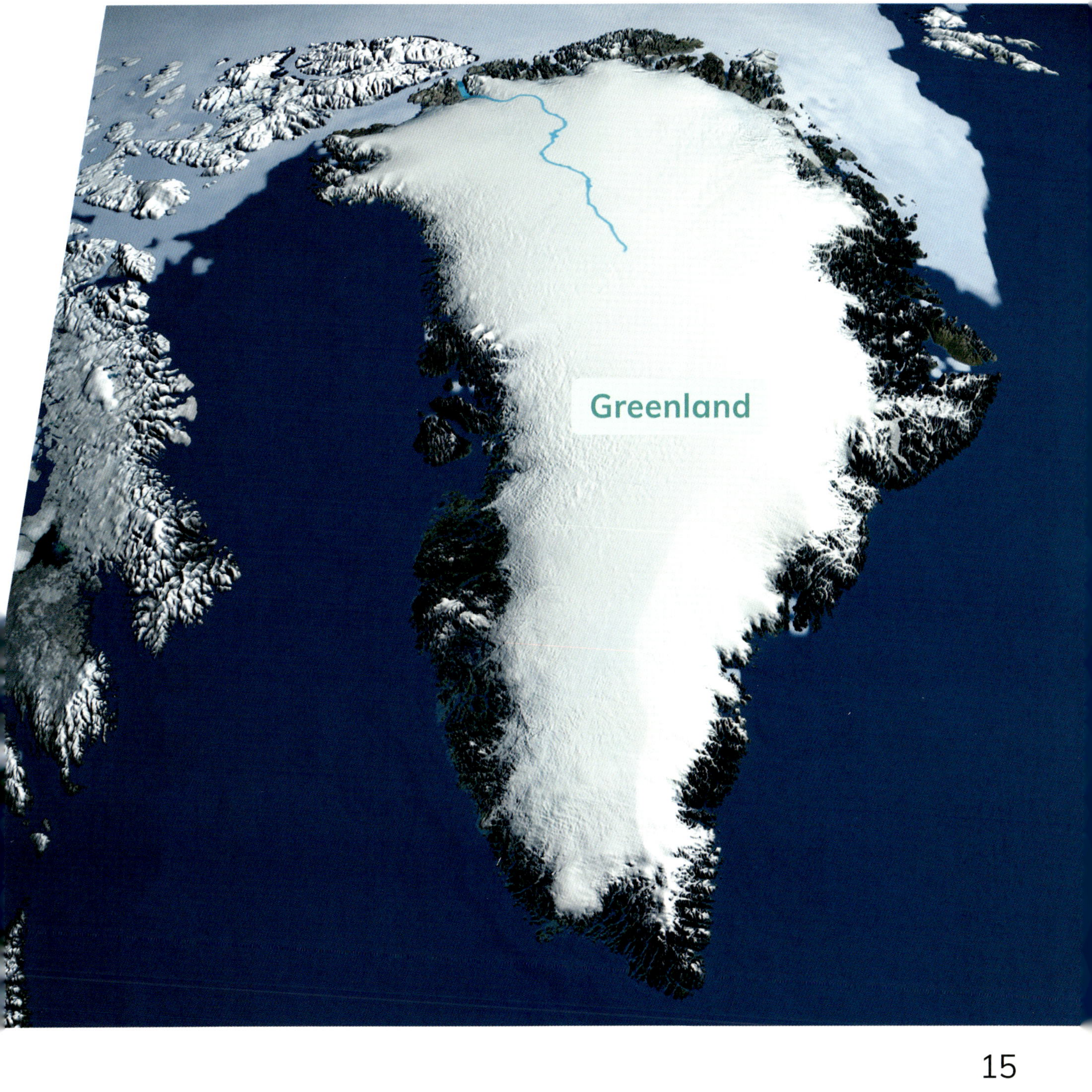
Greenland

Some islands are very small. The smallest island people live on is in the state of New York. It has one house and one tree. Water nearly touches the house's front door. People must take a boat to get to the island.

WHO LIVES ON AN ISLAND?

Many animals and plants live on islands. Flowers and grasses grow. Trees grow tall on some islands. Birds, snakes, and other animals live on islands.

Sometimes people live there too. Would you live on an island?

Make an Island

With just a few materials, you can make an island of your own!

You'll Need:

- brown modeling clay
- baking pan
- water
- blue food coloring
- small plastic toys

Instructions:

1. Shape the modeling clay into an island. Your island can be big or small. It can be flat or have mountains.

2. Press the clay onto the bottom of the baking pan.

3. Pour water around your island. Do not cover it all the way.

4. Add a few drops of blue food coloring to the water.

5. Decorate your island with plastic toys. You can add trees, roads, cars, animals, or houses.

Glossary

coast (KOHST)—land next to an ocean or sea

erode (i-ROHD)—to wear away

iceberg (EYESS-berg)—an ice mass that has broken away from a sheet of ice

landform (LAND-form)—a natural feature of the land

mainland (MAYN-land)—land that makes up the main part of a continent; a continent is one of Earth's seven large land masses

volcano (vol-KAY-noh)—an opening in Earth's surface that sometimes sends out hot lava, steam, and ash

Read More

Allman, Barbara. *Islands*. New York: Enslow Publishing, 2020.

Kalman, Bobbie. *What Are Landforms?* New York: Crabtree Publishing, 2018.

Tanumihardja, Pat. *Amazing Islands Around the World*. North Mankato, MN: Capstone, 2019.

Internet Sites

Ducksters: Island Geography
https://www.ducksters.com/geography/islands.php

Missouri Botanical Garden: Shoreline Topics
http://mbgnet.net/salt/sandy/index.htm

Index